# Idlewild and Woodland Park, Michigan

*"An African American Remembers"*

Published by
Run With It
Grand Rapids, Michigan

Book Concept and Design by
Rose Hammond

Book Cover Design by
Rose Hammond

Copyright @ by Rose L. Hammond
All Rights Reserved
Printed in the United States of America

No parts of this book may be reproduced or transmitted in any form or by any other means, electronic, graphic, optical, or mechanical, including photocopying, recording, taping, filming, electronic data, internet, packets, or by any other information storage or retrieval method without prior written permission of the author.

Idlewild and Woodland Park, Michigan
"An African American Remembers" by Rose L. Hammond

    1. Idlewild and Woodland Park, Michigan - Interviews
    2. Idlewild and Woodland Park, Michigan - Portraits

Non-fiction publication

LCCN 2008928279

ISBN 978-0-615-21722-2

For information contact: Rose L. Hammond at (616) 581-3149
E-mail address rdarlene22@hotmail.com.

The author gratefully acknowledges permission for the following quotes: Page 34 "Big Gray Pike" & advertisement (Ella Foster author, copyrighted 1924) and page 80 "If You're Tired" (Ella Foster author).

Marion Arthur and Ella Arthur pictures provided by Mary Thomasson and Hazel Johnson.

# I dedicate

this book to my father and mother, Melvin and Louise Hammond who loved us unconditionally and encouraged my siblings and me to follow life's course which also may include some dreams. My parents have both since passed on to a higher being and our family miss them both. I encourage everyone to cherish the time that we have with our parents and support them through their life's journey which could include illness.

I have learned that dreams may not take the course we might expect. Be patient, believe in yourself, and your dreams will come true.

I love both of you,


Your daughter, Rose

# Idlewild and Woodland Park, Michigan

*"An African American Remembers"*

*Interviews and photographs by*

Rose Hammond

*Edited by*

Dr. Veneese V. Chandler
Marcy J. Rosen

# acknowledgements

**Without** my family's investment, support, insight and unconditional love this book would not have come to completion. I lovingly thank my immediate family members: Melvin and Gladys Hammond, Fred Hammond, Larry and Lori Hammond, Doneall Hammond, Aunt Lillie Hammond, Uncle Eugene Hammond, Marie Harris, my beautiful mother-in-law, and most notedly my son, Astin Martin.

I would like to acknowledge my investors: Lisa Childs, LaDeidra Gais-Hughes, Marcy J. Rosen, Dr. Veneese V. Chandler, Theda Fields, Dawnn Burrell, Grand Rapids Community Access Center (GRTV), Wyoming Community Access Center (WCTV), and the rest of the production crew.

I would like to thank all of the residents and resorters of Idlewild and Woodland Park, Michigan for their historical inspiration and interviews for this project.

Original of the plat maps for Idlewild are recorded at the Lake County Register of Deeds. The original plat maps for Woodland Park are recorded at the Newaygo County Register of Deeds.

# introduction

There are **numerous moments** in African American history, but this is the story of a moment that still exists. This is the story of Idlewild and Woodland Park, Michigan.

Idlewild was founded in the 1900's and recognized as a place for African Americans to resort. During the second decade of the 20th century a small, yet clearly distinguishable African American middle class largely composed of professionals and small businessmen and women had been established in several urban centers. The Idlewild Resort Company (IRC) organized its first excursion to attract middle class African American professionals from Detroit, Chicago, Indiana, and other mid-western cities to tour the rustic community. During their visits plats were sold. A few years later, the same strategy was used to attract African Americans to Woodland Park.

This book is a historical compilation of interviews conducted during the summers of 1994 and 1995 with some of the few original owners or families still holding property ownership. The reader will see some of the individuals whose families helped to create these two beautiful communities by telling their family's history. Idlewild and Woodland Park, Michigan "An African American Remembers" will let the reader travel in time from the 1900's.

**Note . . . .** The State of Michigan is helping to revitalize Idlewild, already registered as an historical area.

# Idlewild and Woodland Park, Michigan

*"An African American Remembers"*

# narrator

This is a chronicle of memories. Memories of a time and place that changed a race of people; a nation.

This is a chronicle of memories of what started out as a dim chance but also seemed like a bright future.

This is a chronicle of two of the country's most popular and well-known African American resorts; Idlewild and Woodland Park, Michigan.

# introduction

### Reva branch

They got this idea of turning it into a Black resort because the Blacks had no where to go.

### Lillian jones

We had this little place up here, so we borrowed this tent and we lived in the tent.

### John slade

At the longest the season is from Memorial Day to Labor Day.

### Sonny roxborough

But it didn't last. 1970, I can remember it just as that bell. . . . the bottom fell out, "boom," it's all over. My business dropped half.

### Steve jones

If you could see some of the flyers that have been collected you will see how they used beautiful phrases like, "Be your own landlord." "Own something that nobody can take away from you." Those were the kinds of things that made African Americans, Negroes as we were called then, buy property in Idlewild and then later here in Woodland Park.

# narrator

It is the middle of June 1917 and another hot day in Chicago. Daddy is in the living room and mama is in the kitchen when a knock is at the door.

As I peek around the corner from my bedroom there stands a tall, thin, well-dressed Colored man. He talks like he is an insurance salesman, but he is not.

He is talking about this land in upper Michigan. He says the land is being developed as a resort for Colored people.

The land is described as having the most beautiful lakes and tranquil sounds of nature. A place to get away for the summer, a utopia.

Most of the plats measure 25x100, cost $35 each, $6 down, and $1 a week.

$35 each $6 down $1 a week

## chapter i

Our family lived in White Cloud, Michigan and they got this idea of turning it into a Black resort because the **Blacks had no where to go.**

What I remember going up there. . . . grandpa and grandma used to go up in the summertime and my brother would go along. But, it was always manned so that the people could come up and see their property. If they came up and they did not like their lot particularly, they were allowed to trade it for one that they did like because most people bought them sight unseen.

**Reva** *branch freeman*

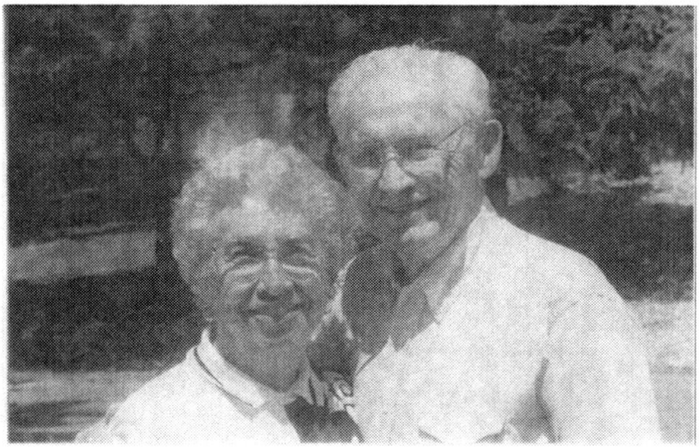

# Map of Idlewild, Lake Co., Mich.

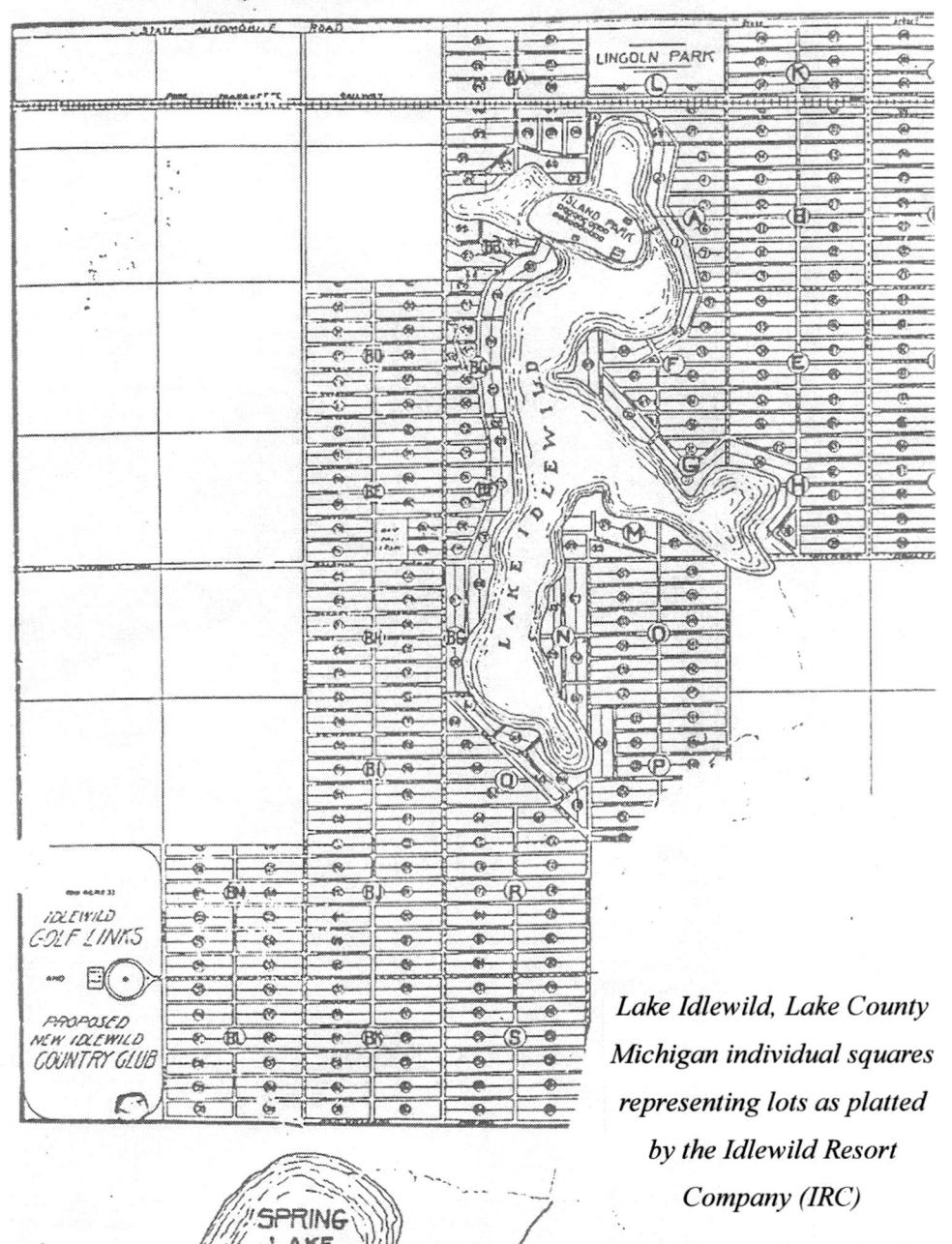

Lake Idlewild, Lake County Michigan individual squares representing lots as platted by the Idlewild Resort Company (IRC)

IDLEWILD RESORT CO.

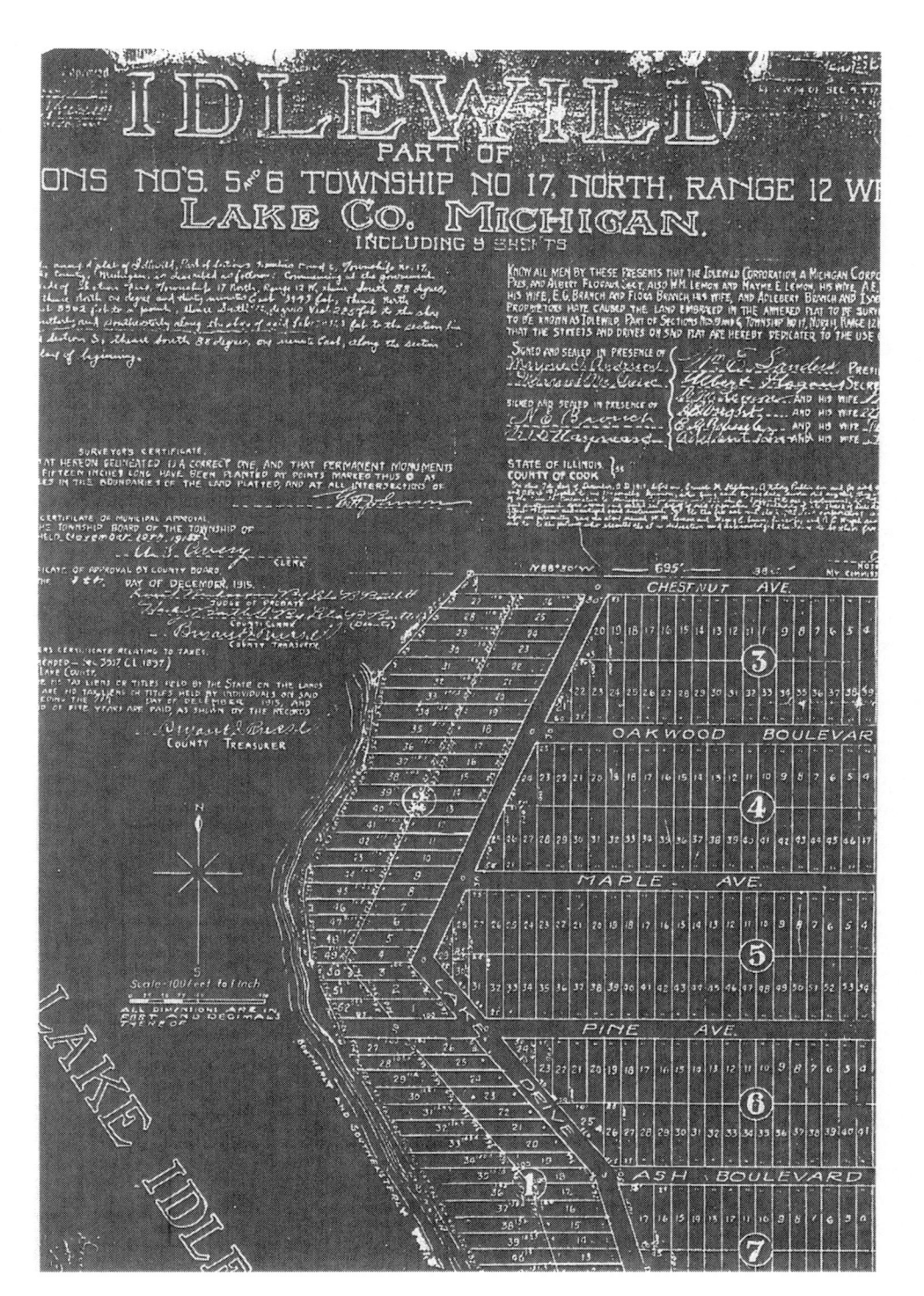

*Plat map of Idlewild, Michigan*

20-21

My family first came to Idlewild in 1932. It was publicized in the Chicago Public Defender. It seem like to me people were migrating from Chicago more than any other place and when they got up here they found out it wasn't farmland, it was mostly sand. A lot of them went back, some of them did stay.

**Charles** *sonny roxborough*

I know that the Branches perhaps or their representatives, anyway, went all over the country and approached people in terms of this being a beautiful spot for Black people. Of course, we weren't Black then, we were Negroes (Colored). You bought lots land unseen and that was all over the world not just in the United States where this approach was used. My people who are from Kansas City, Kansas bought lots up here before we even came to Idlewild. And, so it was many years before we actually owned property where there was a residence. It was described as a utopia.

**Emma** *jean clark*

*Main beach in Idlewild Michigan*

# Woodland Park was

first founded by a man named Marion Arthur and his wife, Ella. They were African Americans. They both had been instrumental in working with the Idlewild Resort Company in Idlewild. He in fact had been a salesmen . . . . one of their chief salesman as I understood and that as the property in Idlewild was becoming scarce, Mr. Arthur went looking for property that he could emulate Idlewild and that is when he founded **Woodland Park.**

**One** thing that I have learned is that Ella Arthur was a master at marketing. That anytime you see any of the pictures that we have, most of these pictures were actually postcards. What she would do is anytime somebody caught a big fish, anytime somebody built a house, anytime somebody did something monumental, Mrs. Arthur would take a picture of it and have it made into a post card.

So, then when that person sent the postcard or when anybody came up and bought a postcard or received the postcard or sent it across the country then other African Americans saw what these African Americans were doing in Northern Michigan. That is what created the whole fervor.

**Steve** *jones*

# WOODLAND PARK

PART OF SECTIONS 4 AND 9
TOWNSHIP NO 15 NORTH RANGE 13 WEST.
NEWAYGO, CO. MICH.
Including 12 Sheets.

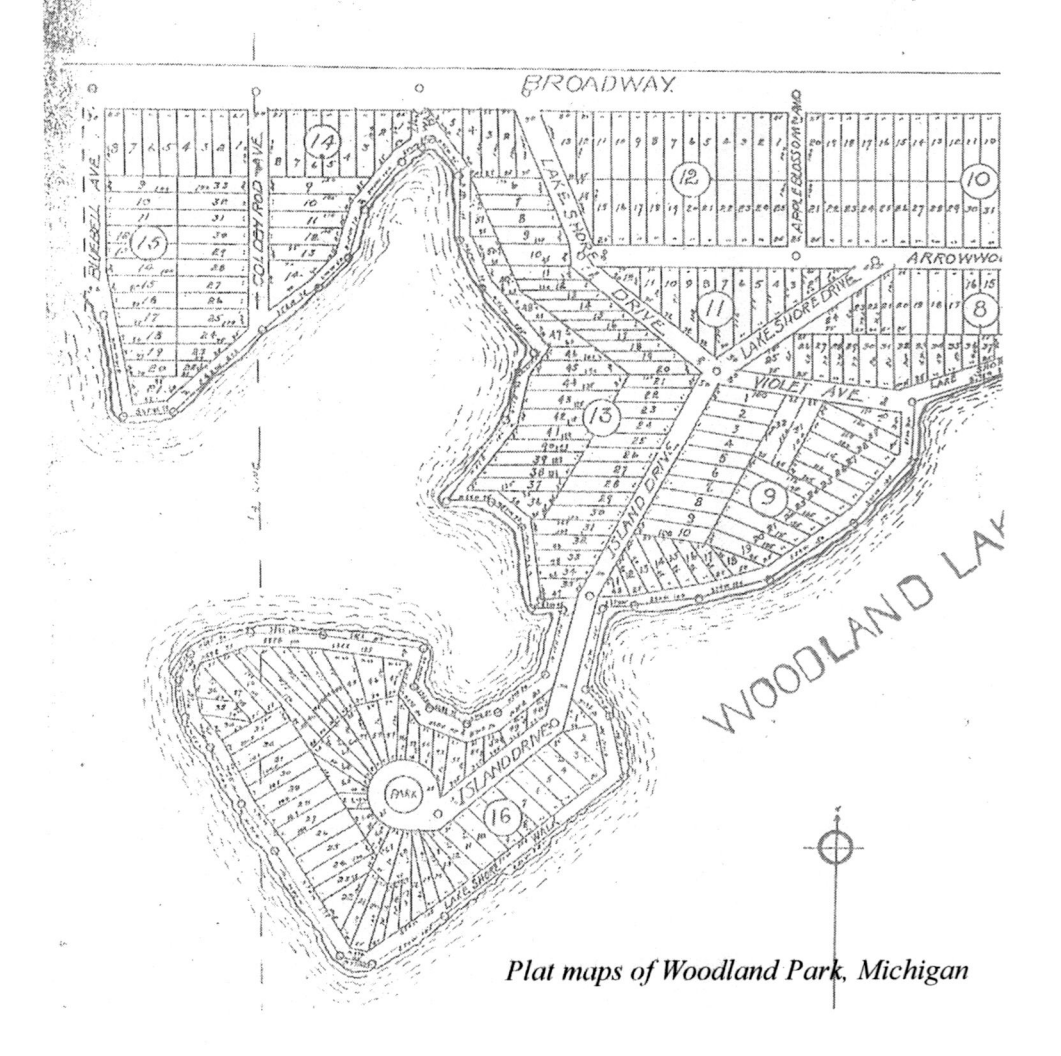

*Plat maps of Woodland Park, Michigan*

# WOODLAND PARK

SHEET NO 1
NE¼ OF NE¼ SEC
T15N. R13W

## PART OF SECTIONS 4 AND 9, TOWNSHIP NO 15 NORTH RANGE 13W
### NEWAYGO CO. MICH.
INCLUDING 12 SHEETS

KNOW ALL MEN BY THESE PRESENTS THAT I, WILBUR M. LEMON, Trustee, have caused the land embraced in the annexed plat to be Surveyed laid out and platted, to be known as WOODLAND PARK, part of Sections 4 and 9 Township No 15 North, Range 13 West, Newaygo County Michigan, and that the streets and avenues on said plat are hereby dedicated to the use of the public, for street and highway purposes only

_Wilbur M. Lemon_ (Seal)
Trustee

Signed and sealed in presence of:

STATE of Illinois } SS.
County of Cook }

On this 15th day of May A.D. 1921, before me _____ a Notary Public in and for said county appeared Wilbur M. Lemon, known to me to be the same person who executed the above dedication and acknowledged the same to be his free act and deed as such Trustee.

_____ Notary Public

My Commission expires _____

Surveyors Certificate: I hereby certify that the...
[illegible certification text]

Certificate of Municipal Approval: This plat...

Certificate of approval by County board...

County Treasurer's Certificate Relating to...

---

BROADWAY

[Plat map showing blocks numbered 1, 2, 3, 4, 5, 6 with lots, streets including ARROWWOOD BOULEVARD, COTTONWOOD AVE, LAKE SHORE DRIVE, ROSEWAY, EASTERN BELL AVE, and WOODLAND LAKE]

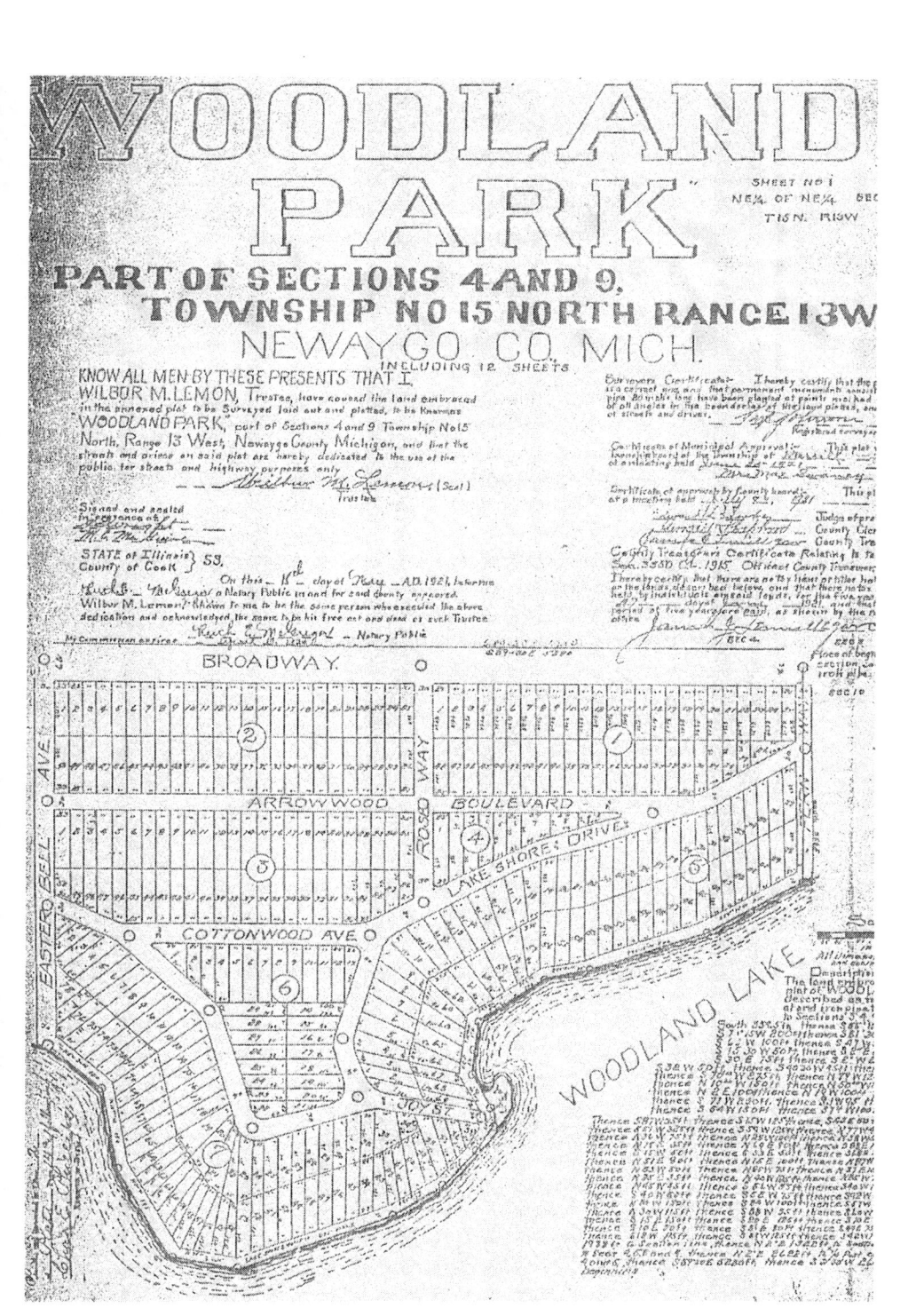

WOODLAND LAKE

I believe my parents first went there to vacation in 1925. That was about four years before the depression.

At that time they had lived quite well. There was a lady.... I don't know if my folks knew this lady personally or if mother just happen to meet her somewhere. But she was telling her, my mother, about the summer resort in Woodland Park that catered more or less to Black people.

At that time, it was rather unusual because there were not too many places for Black people to go. When my mother told my dad about it, he was so excited that there was some place that they could go and relax and vacation like other people did. So, they made plans to come to Woodland Park.

Virginia *proctor*

"They built a home in 1926"

They had little booklets like this with pictures of the hotel and all that they used for publicity. They had postcards that they used to send out giving the rates at the hotel and costs for the breakfast. Breakfast was 30 cents back then. The advertisement tells you about the smooth, wide trails through beauty lands, boulevard highways, well marked for safety and convenience, fascinating landscapes over breezy hills, etc. They have a picture of the inside of the clubhouse. They have a boat livery. They have pictures of people in their swimming suits. That is the kind of things used for advertising. My family is from North Carolina and we first came to Woodland Park in 1921.

**Mary** *thomasson*   **Hazel** *johnson*

**Marion** *arthur*

Founder of Woodland Park, Mi

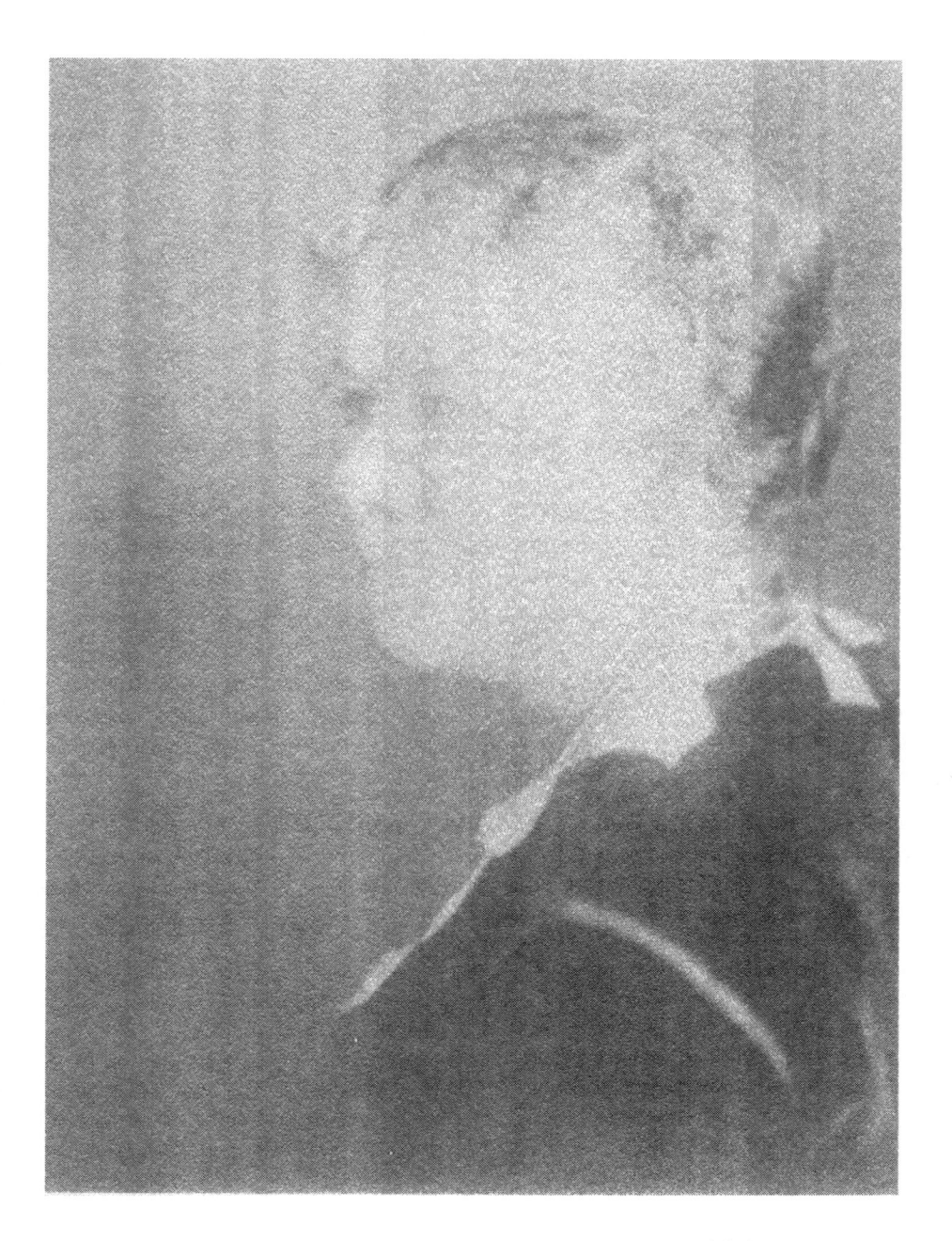

# Ella *arthur*

Cofounder of Woodland Park, Mi

# narrator

Wherever you go, it seems, you are hearing and seeing the various means of advertising of the two resorts.

I picked up a postcard the other day and it reads as follows:

*"A big, gray pike came swimming by and he was old and gay and sly. Summer day's he spent in watching men with great intent, as over their rods and reels they bent. He dived and plunged with all his might. His struggle was a losing fight that the anglers hook and line held fast. The pike had met his match at last. 'Twas then the man's reel turned to laugh, the day that pike could net and gaff and landed him a nine pound fish, fit for a king is such a dish." (Ella Foster author, copyrighted 1924).*

Every summer we start to see more and more people of influence purchasing or visiting the two resorts. Today, I saw Dr. Daniel Hale, a heart surgeon, W.E.B. DuBois, an intellectual, and Madam C. J. Walker. She is in the hair care business. It is a migration of sort which began the building of two communities.

# building a community a migration

## chapter ii

My grandfather, Charles W. Chestnutt, had a court reporting agency in Cleveland, Ohio and since the courts were not in operation in the summer, he was free to travel and come up here. First they would take the lake boat from Cleveland to Detroit and then after that they would drive. See, the boats would take cars. They were auto ferries. They would take cars as well as passengers. That's the way we did it for many years. Well, the trip would take . . . . the boat was an overnight boat. It would sail from Cleveland around 11:30 to midnight, and we would be in Detroit in time for breakfast. Then you would drive up. Of course, from Detroit it's about 200 miles and well, the roads were not terribly good, but you could make it by, I guess, 6:00 or 7:00 o'clock.

Although there was a great deal of the island, Williams Island across the lake was the center of Idlewild and it was built up. On that island there was the clubhouse which was a large structure at this end . . . . at the east end. They served meals there and they would have music and dancing and public meetings and things. It was a gathering place.

Also, on the island there was a *store* which also included a *gas station.* There was one pump. They had groceries, ice and the *post office.* It was run by a family named Elsner. There was also in the middle of the island a *barber shop* for the summer run by a man named Day. Then, at the other end of the island . . . . at the west end of the island there was a *three-story hotel.* I don't know how many rooms it had in it, but they were single rooms for visitors. Then on the back side of the island there were individual *small cabins* which people, tourists, or visitors could rent if they didn't want to stay in the hotel. Then also at the west end of the island was a *night club* called the Purple Palace and all of these things were in full swing during the summer. In the 20's there were not a great many cars. People walked and would walk around on the road with flashlights to attend something on the island and then you would walk back. Some people would row across the lake.

**John** slade

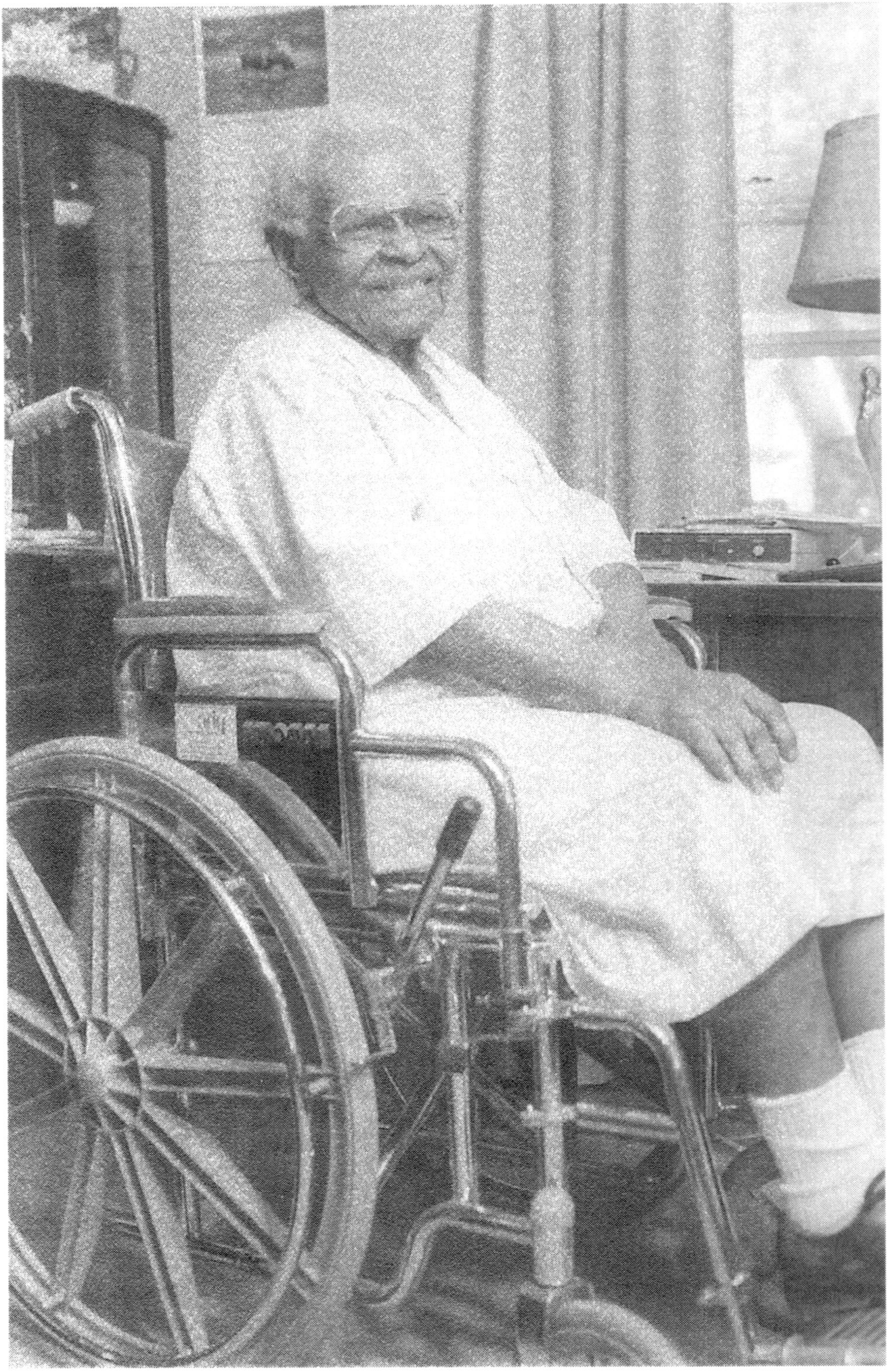

I didn't see Woodland Park on the picture, but we saw Idlewild on the picture on the screen at Dudley's Theatre in Detroit. Years ago that was on Gratiot Avenue and we seen the picture of Woodland Park . . . . I mean of Idlewild.  It flashed on the screen.  It was nothing but woods, dense woods. **We had to cut** through to get to our land. It was just nothing.  It was nothing but woods just like those woods that are coming through now. We are from **Battle Creek, Michigan.**

**Lillian** *jones*

You see, the only thing we had to build with was **tough lumber.** It was that old hard oak lumber. Houses were built out of hard oak. We first came to visit the lots. My husband could not find work because of the **depression** so we knew we had this little place up here. We borrowed some hand tools and lived in a tent for a couple of years until we built us a little house. They started to working on the roads soon. I was the **Township Clerk.** They never had nobody but White men working on the roads and I got our men jobs on the roads. . . .on the County roads. So, that's how they commenced to start working here.

**Lillian** *jones*

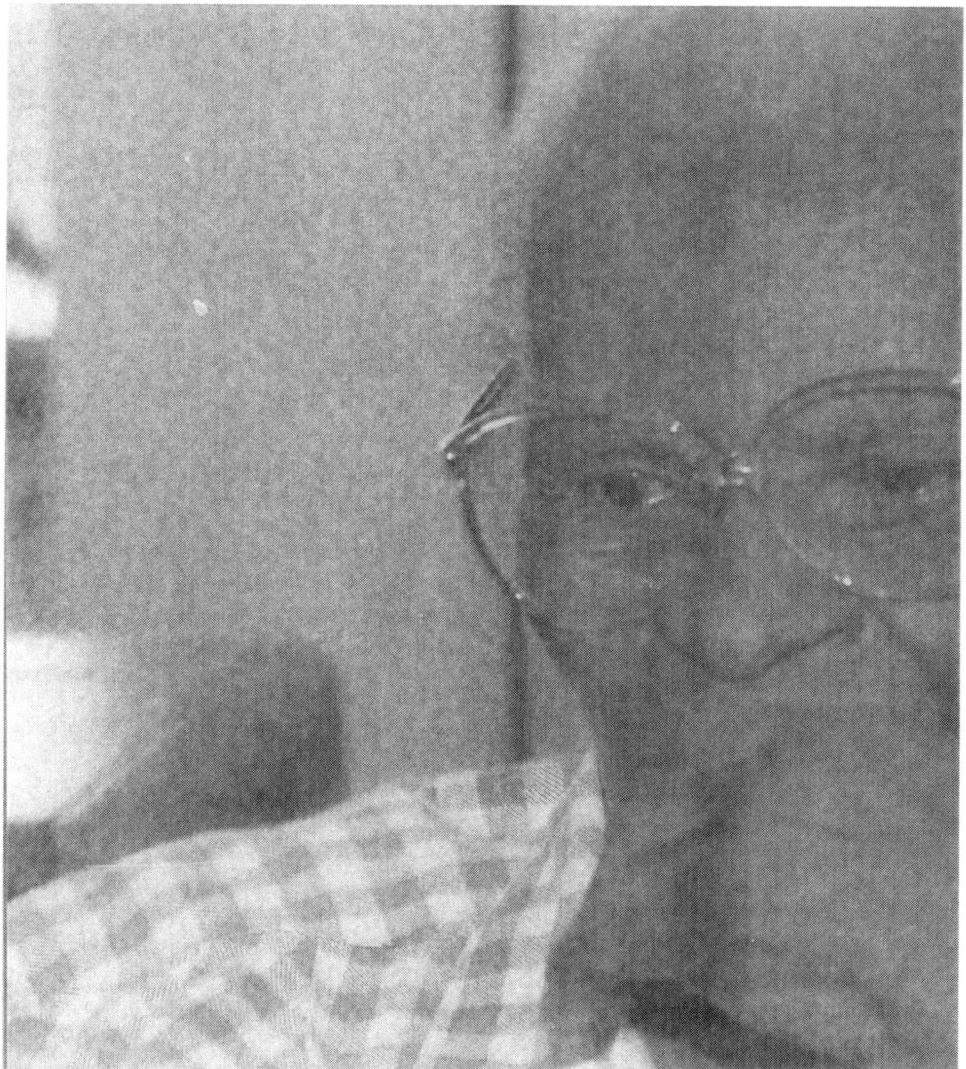

I saw that cottage on stilts. It was awful. My Auntie went up with Mrs. Wilson to look at it. It sat on the lake. And down behind we saw this thing almost falling down and looked like, well. Auntie would holler down and we both said, "No, we don't want that thing." She said, "You come up and look." I had to climb up the hill to get up to the door and she had walked around to the front. I walked around there and there was that beautiful sunset.

And I said, "My God, let's buy the sunset." That's what we bought. Most of the cottages at that time were rafters inside. The walls weren't even finished. Everything was an outhouse. But, people rented.

**Gladys** *chipchase*

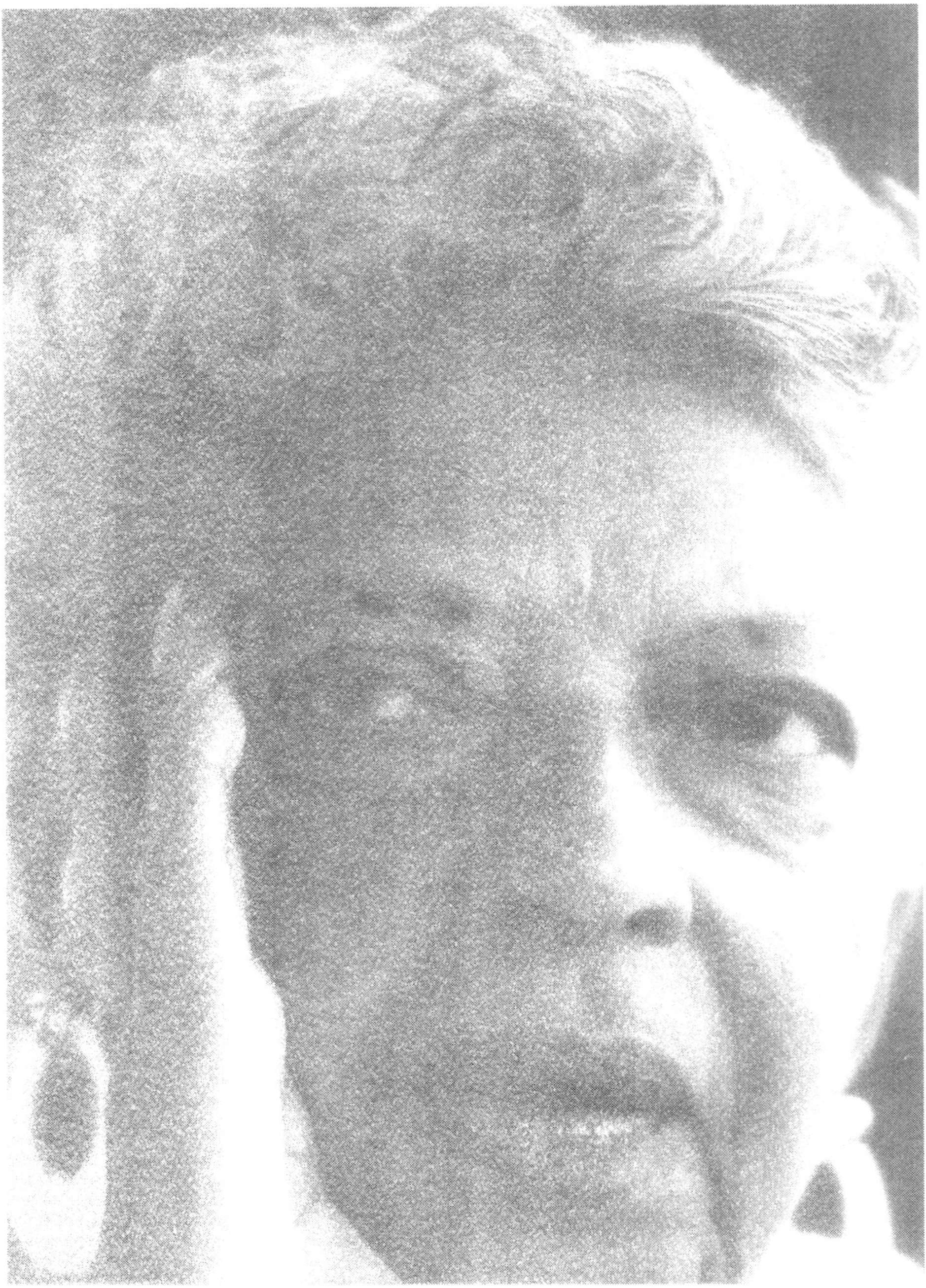

You see, at that time, you *could not* go to the hotel. Now that's when I came along which is somewhat after my friend Gladys. You still *could not* go to the show. You *could not* eat in the restaurants. All they wanted was your money when you went to the grocery store here in town.

So, a lot of the things you see, the doors, a lot of the windows and all those kind of things, they brought up from *Chicago* and then Dad had a builder here who did the actual contract work.

**Ann** *hawkins*

*I never realized what was going on until I was older so to speak. But, we always had to carry lunch. We* **could not** *stop and go into restaurants. The same thing existed all over the United States but to a lesser degree here because we're talking about fewer people. But, certainly until the 50's, you know, things didn't level off.*

**Emma** *jean clark*

*"You always had to carry lunch with you"*

This was basically an area, at that particular time, wasn't attractive to Whites. There were other areas further north, 70 or 80 miles, in the Traverse City area where you had places that individuals, industrialists and people who were coming in from Chicago and various places, they were building and settling there.

When you first started coming here it was just a village of tents, no permanent residences. They lived in tents until they were able to build homes and cottages. You see at the time I started to come up here in the 50's it was the **Flamingo** and then they had the **Paradise Club** and the **El Morocco.** Then, we had **Pearl's Bar,** the **Roxanna Bar, Eagle's Nest,** and several **restaurants** were around at that time. That was where you had the Las Vegas type floor shows complete with chorus line, comedians, the musicians. The Braggs out of Saginaw was the ones that used to promote the shows at the Paradise Club.

**John** *meeks*

"Everything was top of the line"

*I heard about Woodland Park in 1950 when my husband described the places they were visiting when he was growing up. But, I never saw Woodland Park until 1954 when I came here from* **Savannah, Georgia***. We had kids and we wanted to make sure they had a place like this to go to in the summer. It turned out that my next door neighbor who is Mrs. Thomasson's father, was selling lots in Woodland Park and so we bought the lots. But, in the meantime, we would spend the summer at the* **Royal Breeze Hotel** *until we purchased our own place. The* **cottages** *were not modern, it had* **a bed, a stove** *and you would bring everything else that you needed. What I remember is that every morning we would go fishing. They would wake us up at 5:00 a.m. before it was time to start breakfast in the hotel. We would paddle up the east lake and fish until the bell rung. That simply meant that it was time for Mr. Harris to come in to start helping to get breakfast ready at the hotel. Mr. Harris would fry your fish for you. No matter where you lived in "the Park". . . .that's what we called Woodland Park. . . .if you were a teenager, you headed for the porch of the Royal Breeze Hotel. You see, it was not just a hotel, it was a place to sleep. You must remember, the Royal Breeze Hotel was a hotel in the days when it was not popular to go to the Holiday Inn. To me you were not welcome with a whole bunch of kids at the Holiday Inn. A flashlight was necessary because you would not see much more than a few feet ahead of you, if you were lucky. All the kids traveled the roads at night walking with their lights. You could be looking and you would see just lights bobbing up and down the road and there would be bunch of kids going home at night from having been at the hotel.*

**Kathy** Miles

*Steps to the Royal Breeze Hotel from Woodland Park lake side.*

Well, all the businesses were **Black owned.** *It's funny that when I look back now. . . . I mean some people just took their kitchen in their house and turned it into a restaurant. When I think that those situations created entrepreneurs, that you found a lot of people just trying eagerly to get in business. They would start in their own homes sometimes. But, the* **first** *businesses in Woodland Park. . . .when the Arthurs came here there was a* **logging house** *and Mr. Arthur was going to tear it down but once they got the outer coating off they saw that it was made of good hard wood, so he decided that it was best to just build on to it. And, that is what he created the* **Royal Breeze Hotel** *from. It sat on the bank and people who could remember it will never forget it. The Royal Breeze Hotel was majestic. It sat up there like something out of Egypt.*

*There was another building that was right as you started to go down the what was then a two-track because all these roads were two-track then. It was too good to be torn down so he decorated it and he turned that into the registration place for the hotel. He sold lots out of there.*

*Earlier from some of the first settlers it was used as their church, as their worship place. I think it was used as a funeral parlor for some of the early members that died here. Then, it was later expanded and he turned it into the* **Pine Cone Tavern.** *At the Pine Cone Tavern he had* **gas pumps.**

*Somewhere around probably the late 20's, the early 30's Mattie Keller and Ella Towns came to Woodland Park. They built the* **first store** *in Woodland Park. It was a coffee shop and oil station. And then on the same corner, I don't know how long they kept it, but then they expanded that into a store, a grocery store. Then, they expanded even that. They also built one of the second or third* **hotels** *here, very elegant, a huge building. I mean a humongous building. They called that the* **Calesonia Ranch.**

**Steve** *jones*

54-55

I remember that the Paradise Club was right across the lake. I could hear all the music. In the summertime the cars were almost **bumper-to-bumper** up here. People would sleep in their cars because they could not find enough places to sleep in. They would also go to their friend's home who had a rental to wash up and then enjoy the day. But, all the stars that could get here came up to Idlewild. One after another would rent my cottages.

The first one that rented my cottages was Ruth Brown. From then on all the stars after they saw the place and they were booked at the club, they stayed with me. I have rented to Aretha Franklin, George Kirby, Brook Benton, Bill Doggett, Jerry Butler and the Four Tops. They have all stayed with me, they have all stayed with me.

**Eloise** brewer

*Well, it was like I say, B. Giles owned the Flamingo and I would work there in the mornings, when they first put in the bar which holds about 150 people. After the show time began it was hit the road again to go and pick up other entertainers. So, we would go back and forth to Detroit. The first entertainer I brought up here was in a station wagon with the band, the whole band and all the instruments.*

# Helen *curry*

*"Everything in that Ford"*

*I remember coming to Idlewild to see the "Hidy, Hidy Ho Man" in about* 1942. *My next trip was to see Dinah Washington at the Paradise Club. The wonderful entertainers, the entertainment. I didn't think of it as living here or anything. It was something so different you know. The only chance I'd ever have to see the big entertainers was to come here.* Blacks *could not attend the White night clubs then and we also could not go to any resorts. We* had to *have our own resort and it was just, it was just the only place you could go and* not be insulted.

**Rita** *collins*

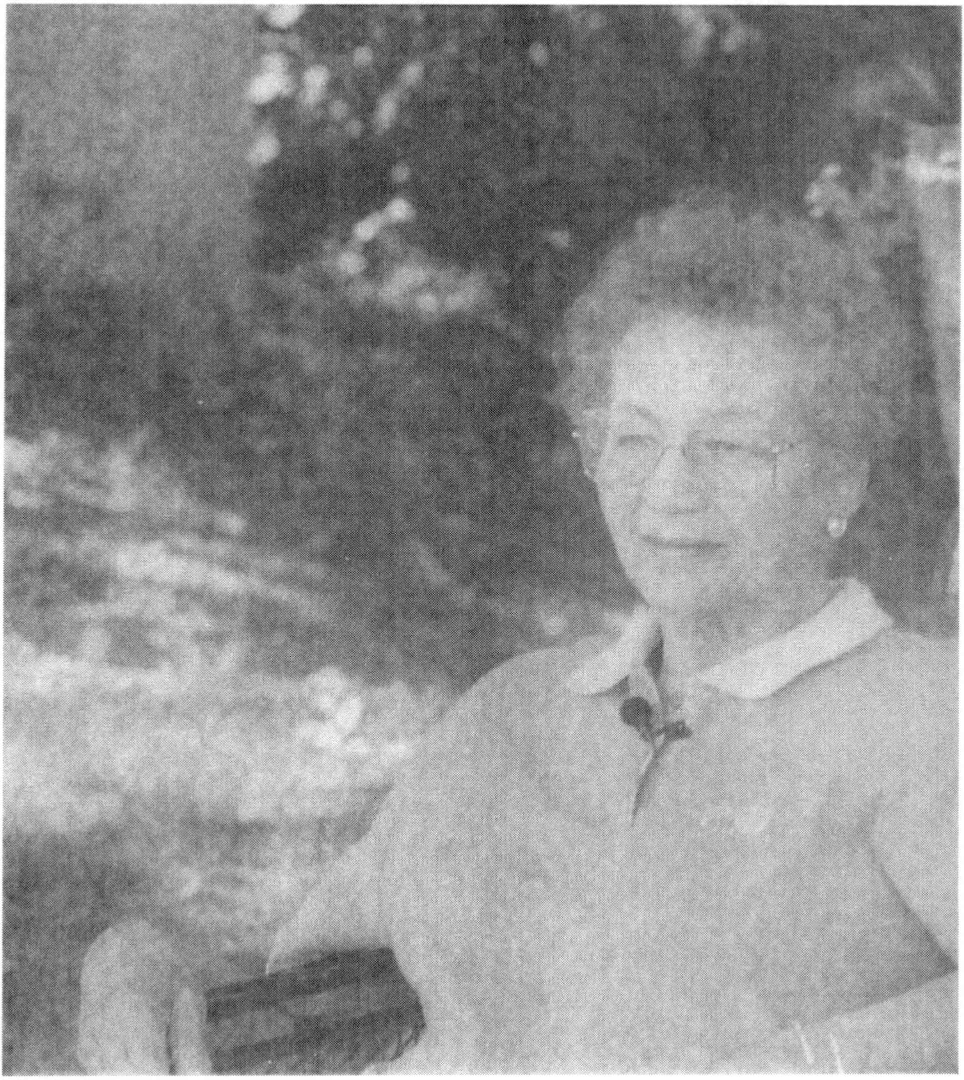

*And we saw all the* **top entertainers** *because they used to say that Idlewild was to the Black entertainer like the Catskills is to the Caucasians. And it was because all of them, majority of them, got their start right here.*

**Ann** hawkins

"But, that era is now over"

# narrator

The newspapers have been talking about the Civil Rights Act. People believe that this will make some things right for Colored people. That it would mean freedom and equality to a Colored society. I don't think the Idlewilders or Woodland Parkers ever thought about the impact that this time in history will have on our two small communities.

Little did anyone realize that with the passing of the Civil Rights Act, some people might not return to Idlewild or Woodland Park in the summer.

As one looks around more and more cottages, motels and even the night clubs are less than half full. Idlewild and Woodland Park are starting to feel the impact of the economic growth with the passing of the Civil Rights Act.

Our family has decided that we will not be back to the area of Idlewild or Woodland Park next summer to vacation. This is a sad day. I wonder when I will see the friends that I have met over the years.

# choices

## chapter iii

Well, Idlewild was. . . .it was a summer place, you know, the season is short.  At the longest the season is from Memorial Day to Labor Day and it gets quite cold at the end of August, too.

For businesses up here and people that want to start something up here that's really not long enough. You can't make enough money to do it. The people that had stores and things like that. . . .that was very often a year-round business. And there are year-round residences in Idlewild. But, they would be people that would live on roads away from the lake. . . . streets away from the lake where it is much warmer.

**John** *slade*

I think Idlewild has provided an **opportunity** for many, many people to have a part of the greater life, the better life so to speak. Now, with **desegregation,** many of them were hurt. You take some of the artists that were out here, you know, and I can use again Kansas City as an example. People say, why would you come this distance? And, it's a valid question. But, it shows we have some commitments of love for it all because we can only go a few miles and find the same thing.

**Emma** *jean clark*

66-67

*You see, even though there are other resorts that started, Idlewild just seems to have thrived.*

*Now, people will say, again, Idlewild has died or is dead. Well, that was because people came during the night club era. Some of us were here* **before** *there were any* **night clubs**.

**Ann** *hawkins*

I don't know if you'd say it's outgrowing a place. You just have so many other **choices.** And, it's just like we are today, you might be able to afford to go to Hawaii every year, but then you might say well, I might want to go to the Caribbean. I have a **choice.** Hawaii's beautiful. But, I think I'll **choose** to go to the Caribbean this year. I think it's very much the same. People had **choices,** they could go where they wanted and they **chose** to go. Of course, at that time, Woodland Park, it wasn't that spectacular of a place. It was just a simple, restful, quiet, lake resort. It didn't have everything to offer. It didn't have the night clubs or hotels where you could live and a pool that you could go to. There was so much more to offer in many of the places where they could go.

So, that's what they did.

**Virginia** *proctor*

68-69

"I think they had
# choices"

The *60's* is when things opened up for African Americans. When **you could** start going to Las Vegas, **you could** start taking cruises, **you could** start doing a lot of things. Not that all those doors were always closed, but now you develop the economics of where **you could** afford to do it and you develop the know how, the intelligence to know how to go abroad or how to speak another language, how to do those types. . . .those things are what, to me, helped create the demise of places like Woodland Park and Idlewild. They expected that people were always going to use outhouses, that people were going to always, you know, accept it. You would prefer to go to another place regardless of who owned it if it had inside toilets. So, those kinds of things, I think helped create the demise of Woodland Park and Idlewild.

**Steve** *jones*

People made money at that time, yeah, but we weren't foresighted. We never thought the day would come along when integration would come. They sent men up here from Detroit interviewing me, "What happened, Sonny?" I said, **"Integration."** They jumped. "Integration is beautiful." I say, "Yeah, not for Black business." And, they didn't leave us because it was Black or White. Because we weren't competitive. We were telling people bring your own laundry. That wasn't right. Furnish your own towels because we figured we had a strangle hold. Never did think that the day would come along like it did come when they could go anywhere they wanted for less money.

Never dawned on us. It never dawned that the day would come along. At that time, there was no where else to go but Idlewild. You would have more fun and meet more people up here. You would meet people up here from all walks of life. **1970,** I can remember just as clear as that bell, the bottom fell out just like "boom" it's all over. My business dropped half.

## Charles *sonny roxborough*

# narrator

The year is 1990 and this is indeed a happy day. For the first time I am able to vacation with my family in Idlewild and visit Woodland Park. It is not quite the same as when I left. I see a lot of demise of these two abandoned resorts. As I visit from home, I listen to people tell their memories.

# memories

## chapter iv

*When you arrive to open up your cottage somebody would be down to help you get the door open. We would ask is everything running okay? "How is your pump?" "Is it working?" "Did you remember to bring bread and milk and things you might need?" "Does anybody need anything from the store?" There was just the camaraderie. But, the first time we came we went to the island and stayed at the hotel. There was a hotel on the island and then there was a clubhouse. And then there was that wonderful beach that you could run around on. You could swim in.*

**Kathryn** browning jeffries

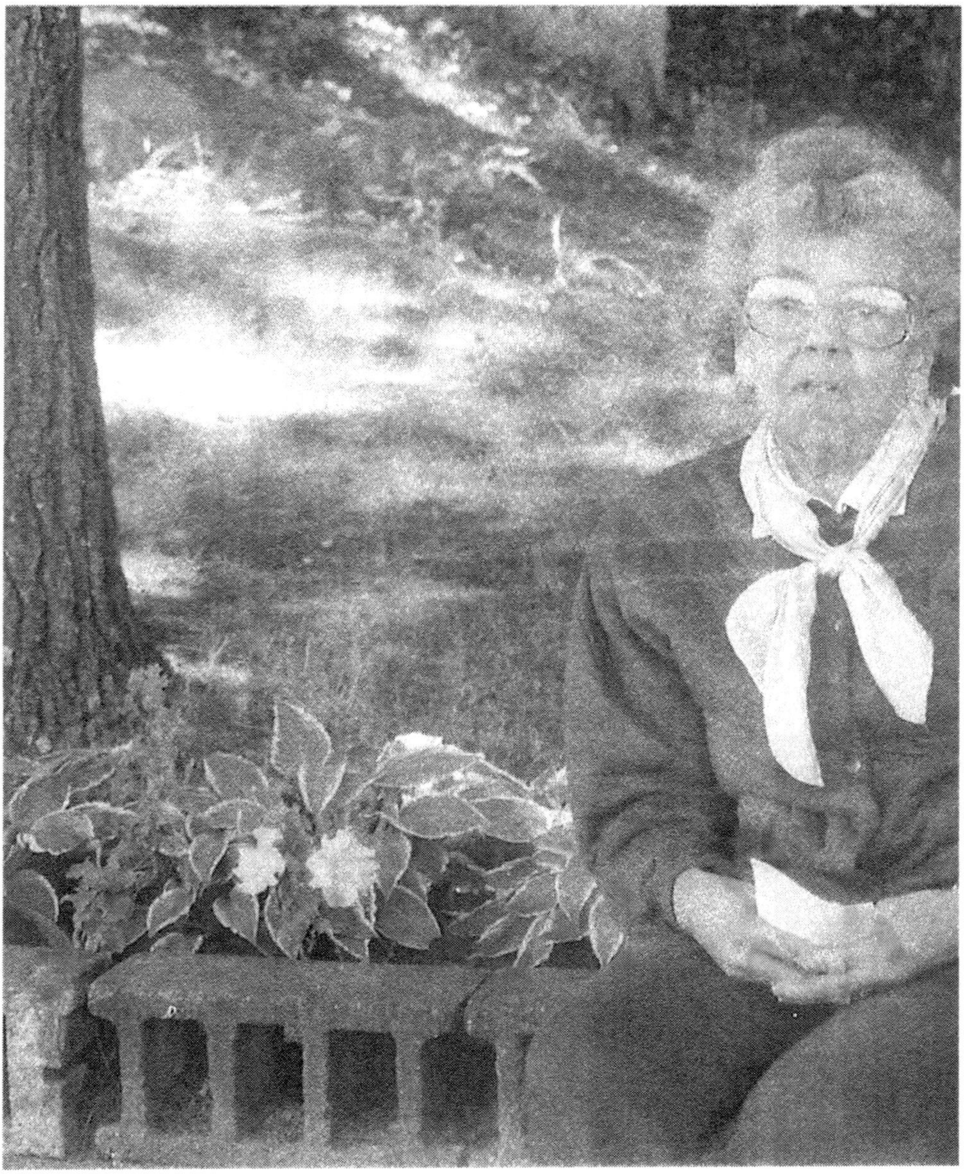

It shows one of the things that we had to **struggle** so hard for, just like **struggling** to get to sit on the bus in the right place. We had to **struggle** for Idlewild because we couldn't go any place else. It's one of our **accomplishments.** That's what Idlewild. . . .one of the things that it means to me. It is one of our great **accomplishments.**

**Rita** *collins*

Idlewild is a place that is still in existence. A place that has gone through a lot of transitions and a lot of notable people have visited here. There has been a lot of things done here in the community to try to preserve some of the area.

**John** *meeks*

I can't think of all the names. . . .entertainers. Well, all your

## State Representatives, your

## Congressmen, singers, movie stars. The

Chipchase cottage was one of those cottages where you had to

come before you left Idlewild.

**Gladys** *chipchase*

I think one thing about it is that we all -- all of us here have made friends from all over the country. I mean, I still have friends that came every year. Now, sometimes, I look back and they are families that you read about and so forth. Those were the people that could come and did come and either built homes or came every year.

**Ann** *hawkins*

I will read this from one of the advertisement for Woodland Park that Ella Arthur, one of the founders of Woodland Park, wrote. *"If you're tired of the city's toil and strife, pack your grip, take children and wife. Hop a train that's bound for the Royal Breeze Hotel, just a place to rest the crowded breeze. Here you will meet your friends from far and near. Joy and laughter brighten days most drear. Not a chance to mope over moments sad, come to Woodland Park and be made glad. Glad that life is coarse in your veins, glad you found a place where God reigns. Glad that there is hope and faith among your race, glad to meet your brother face-to-face."*

**Steve** *jones*

A place where you feel comfortable. Where the neighbors that you've gotten to know are like members of your own family. And, to the people you say at the end of the summer . . . .

*God willing, we'll see you again.*

**Kathryn** *browning jeffries*

CPSIA information can be obtained at www.ICGtesting.com
Printed in the USA
LVOW11s0703110814

398513LV00002B/102/P